Dangerous Fishes of Western Australia

J.B. HUTCHINS

Department of Ichthyology,
Western Australian Museum

ISBN 0 7244 8224 5

Published by the Western Australian Museum,
Francis Street, Perth, Western Australia 6000.

Cover picture: The Red Firefish (Pterois volitans), *perhaps the most spectacular of Western Australia's dangerous fishes. Photograph: J.B. Hutchins.*

Contents

Introduction

Man most fears the sea because of his belief that dangerous animals lurk beneath its surface, waiting to attack those who enter this alien world. While there are many valid reasons to support this attitude — people are bitten, stung, poisoned and even eaten by aquatic animals — it is most probably fear of the unknown, together with news-media sensationalism, that contributes to this anxiety.

People who make their living from an aquatic environment generally recognise and understand the dangers and are capable of taking measures to protect themselves. However, others who use Western Australia's oceans and rivers mainly for recreational purposes are less aware of both the hazards and the need for taking precautions. All divers, anglers, netters, shell collectors and others who use our waterways should be familiar with their more commonly encountered dangerous inhabitants.

This booklet covers the most dangerous group of aquatic animals, the fishes. As well as being a guide to the identification of the more common Western Australian species, it presents information on first aid and suggests preventive measures that should be taken.

At this point it is worth mentioning that members of one group of fishes, the eels, are occasionally mistaken for sea-snakes, and vice versa. This confusion should be clarified, because sea-snakes possess venomous fangs, whereas eels do not. Eels may be easily distinguished by their generally scaleless bodies (the species inhabiting fresh water are an exception) while sea-snakes, like all other snakes, possess prominent scales. In addition, on each side of the head eels have a gill slit, which sea-snakes lack. Sea-snakes are discussed in another Museum booklet, *Dangerous Snakes of Western Australia,* by G.M. Storr.

A few technical terms have been used in the following text. These are illustrated in Fig. 1.

Western Australia occupies a broad belt of latitude, ranging from 35°S on the south coast to 14°S at its northernmost point. It therefore possesses a very diverse fish fauna, totalling some 1400 species. Along the southern coastline dwell the cool water, temperate fishes, while tropical species inhabit the warm waters of North West Cape and farther northwards. The intermediate area (i.e. the west coast) possesses elements of both faunas, as well as a distinctive sub-tropical fauna consisting generally of species found only in Western Australia. This diversity is reflected in the number of dangerous fishes inhabiting our waters, ranging from the fearsome White Pointer Shark to that angler's nuisance, the Common Blowfish. It is, of course, impossible to discuss every species which may be potentially harmful. For instance, there are over 50 species of scorpionfishes in Western Australian waters, all of which possess venomous spines. Therefore the following discussion is restricted to those species most often encountered, whether by casual visitors to our waters or professional fishermen. Most inhabit the marine environment while a few are freshwater dwellers.

There are three categories of dangerous fishes:

- Venomous fishes (those that sting)
- Poisonous fishes (those that poison)
- Aggressive fishes (those that bite)

Fig. 1: Some structures of a typical fish

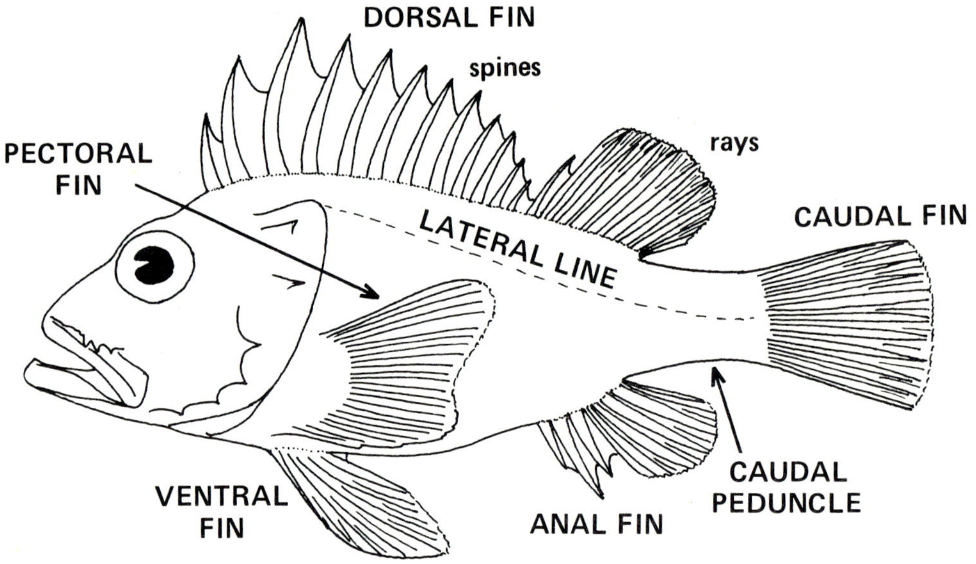

Venomous Fishes

This category contains those species which possess spines capable of introducing toxins into the body on puncturing the victim's flesh. The toxin is stored in venom glands associated with these spines. Release of the toxin is accomplished by a variety of methods. Species belonging to the catfish, scorpionfish, stonefish, stingray and spinefoot families are discussed.

Plate 1: Black Catfish

Plate 2: Glencoe's Catfish

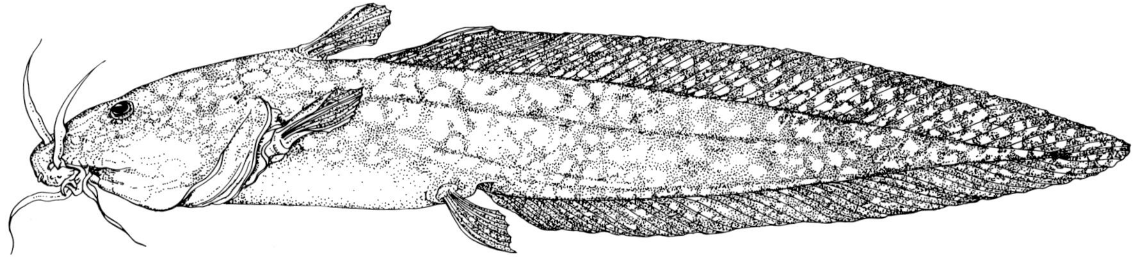

Fig. 2: Cobbler

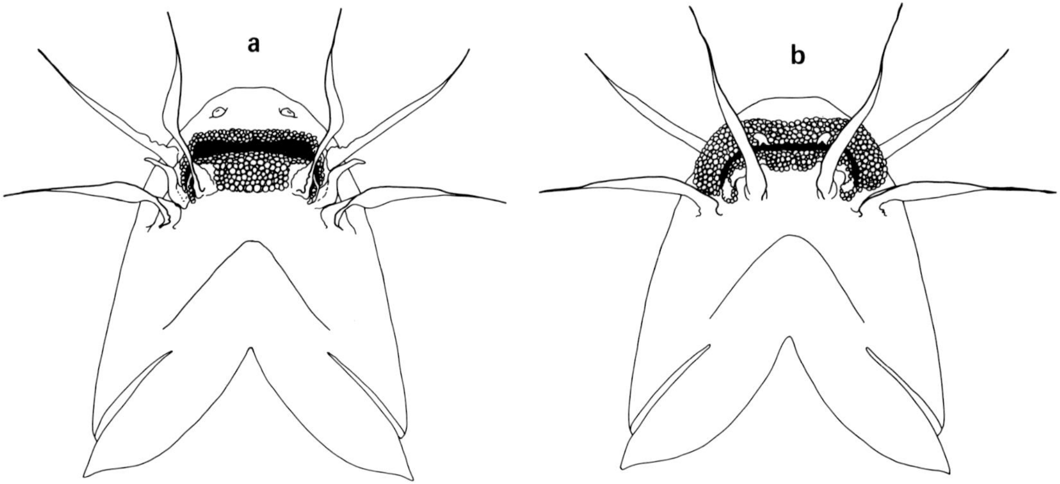

Fig. 3: Bottom view of the heads of (a) the Cobbler and (b) the White-lipped Catfish.

Catfishes

Catfishes may be divided into two groups, the eeltail catfishes and the salmon or forktail catfishes. Both forms possess three venomous spines, one in the leading edge of the dorsal fin and the other two similarly situated in the two pectoral fins. The spines are covered by a thin layer of tissue which contains numerous small venom glands. On puncturing a victim's flesh this tissue is ruptured, releasing the venom. The amount of venom entering the wound depends on the depth of penetration. In addition, the spines are serrated and so tear the flesh on withdrawal. This facilitates absorption of the venom.

Eeltail Catfishes

Cobbler or Estuarine Catfish (Figs 2 and 3a)
Cnidoglanis macrocephalus

The Cobbler is the best known of the venomous fishes found in the south-west of Western Australia. It is common in estuaries such as the Swan River and also inhabits rocky and weedy areas along the coast. It may reach a length of 90 centimetres.

Distribution: southern half of Australia (except Victoria and Tasmania) to the Houtman Abrolhos in Western Australia.

White-lipped Catfish (Fig. 3b)
Paraplotosus albilabris

This species inhabits reefs and weed beds along the west and north-west coasts of Western Australia. It may be distinguished from the Cobbler, which it resembles closely, in the positioning of its nostrils. They are located on the lower surface of the upper lip, whereas the Cobbler's nostrils are found on the lip's upper surface (see Fig. 3). In addition, the White-lipped Catfish lacks the small tentacle that the Cobbler possesses at each corner of its mouth (Fig. 3). The White-lipped Catfish grows to a length of 35 centimetres.

Distribution: Indonesia, New Guinea and the northern half of Australia to Point Peron in Western Australia.

Black Catfish (Plate 1)
Paraplotosus species

This species was discovered in the Dampier Archipelago in 1972. It lives on coral reefs and is easily recognised by the very high first dorsal fin and its overall black colour. This catfish grows to 30 centimetres in length.

Distribution: known only from Western Australia, where its range extends from the

North West Cape area to the Dampier Archipelago.

Freshwater Cobbler (Fig. 4)
Tandanus bostocki

This species inhabits freshwater streams in the south-west of Western Australia. It may be distinguished from the catfish of estuarine regions, the Cobbler, by its more robust and shorter body (see Figs 2 & 4). In addition, the Freshwater Cobbler is only found in fresh water whereas the Cobbler prefers estuarine and marine conditions. It grows to 40 centimetres in length.

Distribution: south-western Western Australia, from the Frankland River (Nornalup) to the Moore River.

Glencoe's Catfish (Plate 2)
Tandanus glencoensis

This northern freshwater species is easily recognised by the absence of a fin along the mid-portion of its back (see Plate 2). It is very abundant in the rivers and streams of the

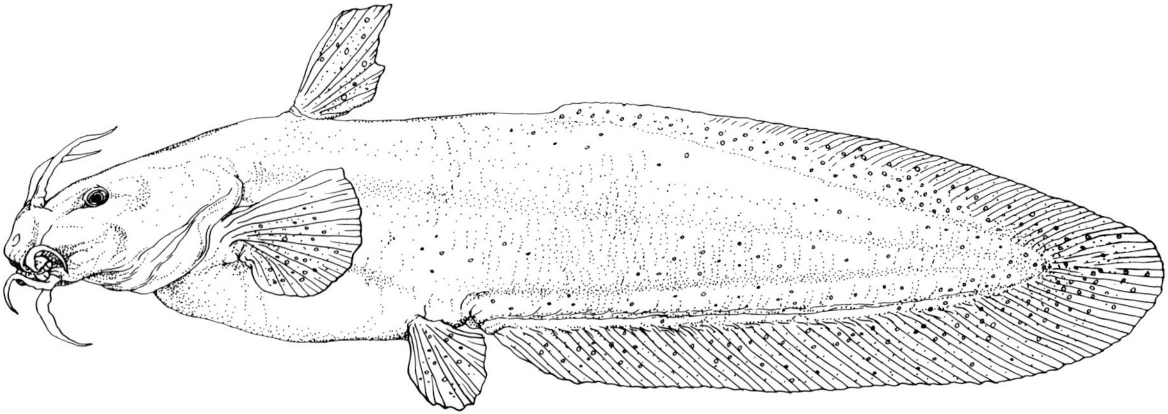

Fig. 4: Freshwater Cobbler

Kimberley and Pilbara regions of Western Australia. This species grows to a length of 20 centimetres.

Distribution: Northern Territory and Western Australia where its southern limit is the Ashburton River.

Long-tailed Catfish
Euristhmus lepturus

This species is most often encountered by prawn sorters on trawlers working in Shark Bay, Exmouth Gulf and Nickol Bay. It is a rather slender fish, growing to 45 centimetres in length.

Distribution: northern half of Australia to Shark Bay in Western Australia.

Striped Catfish (Plate 3)
Plotosus lineatus

This northern coral reef inhabitant is occasionally found as far south as Rottnest Island. It is a schooling species, the juveniles sometimes forming close-packed balls of moving fish (see Plate 3). The yellow stripes on the side of the body are characteristic for small juveniles. These stripes gradually change through white to a silvery colour before fading in the large adult stage. This species grows to a length of 30 centimetres.

Distribution: Indo-West Pacific. Rottnest Island is its southern limit in Western Australia.

Plate 3: Striped Catfish

Salmon or Forktail Catfishes

Giant Salmon Catfish (Fig. 5)
Arius thalassinus

This species is relatively common along the north-western coastline of Western Australia, where it is found both in estuaries and at sea. It grows to a length of 90 centimetres.

Distribution: northern Australia to the Houtman Abrolhos in Western Australia. Also found in the Indonesian-Malaysian region.

Smaller Salmon Catfish
Arius australis

This catfish is relatively common in freshwater regions of Western Australia's northern coastal rivers. As with other species of salmon catfishes, the venomous spines are very strong and sharp, and can penetrate the rubber soles of normal footwear with ease. However in such cases, the venom-carrying tissue may remain on the external side of the footwear, thus preventing the venom from entering the wound. This species grows to 45 centimetres in length.

Distribution: northern half of Australia to the Pilbara region of Western Australia.

Fig. 5: Giant Salmon Catfish

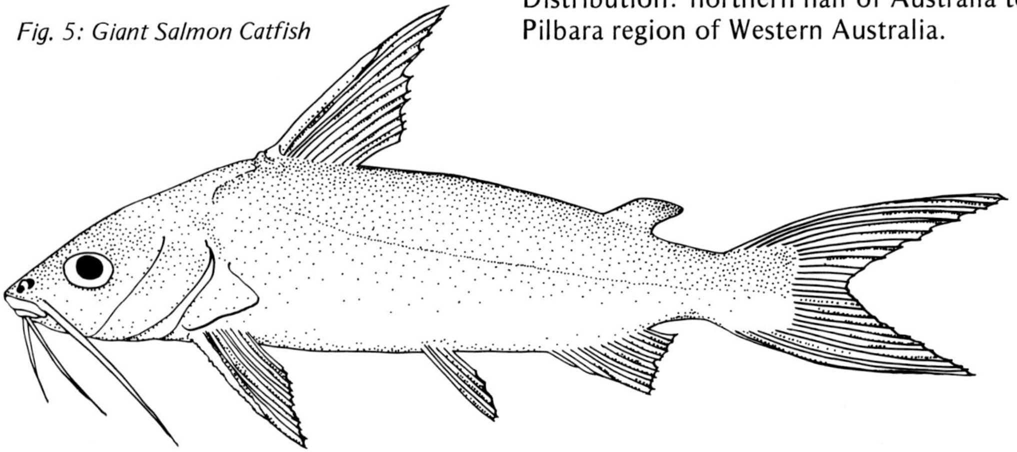

Prevention, Symptoms and First Aid

Prevention

- Never handle a catfish while it is alive. A hard blow to the top of its head should kill the fish. It can then be safely handled, provided care is taken to keep clear of the spines. Both recently killed and frozen specimens are still venomous.
- When walking through weed beds in shallow water, use footwear and shuffle the feet to frighten off any catfishes that may be nearby.
- When sorting the catch of a net, keep an eye out for any fish with barbels around its mouth.

Symptoms

- Immediate pain which increases in intensity for a short while, and may move up the affected limb. The pain usually is severe for several hours and then reduces in intensity to a dull ache.
- The area around the wound generally becomes swollen.
- Some difficulty in breathing may be experienced.
- Professional fishermen claim that a person stung many times may acquire a tolerance to catfish venom.

First Aid

- Immediately wash the wound in water (salt water if on the ocean) and encourage bleeding.
- Immerse the area in hot water (as hot as bearable) and repeat until pain subsides. The venom is denatured by heat and while relief is almost instantaneous, the pain will return if the limb is removed from the hot water, or as the water cools. One to two hours of repeated hot water treatment may be needed before the pain decreases enough to be bearable. Do not apply hot water to an unconscious patient as scalding may result. For areas that cannot be immersed, apply a towel soaked in hot water.
- A tourniquet should not be used, as incorrect procedure may lead to more dangerous consequences than those resulting from the sting.
- As a precaution against secondary infection, medical assistance should always be obtained.

Scorpionfishes and Stonefishes

The sea along the northern half of Western Australia's coastline contains many species of scorpionfish, while lesser numbers are found along the southern half. Most are secretive, sheltering during the daytime in caves, under coral or in weed beds. Some are showy fishes with bright colour patterns and elongated fins (e.g. the Red Firefish) while others such as the Devil Fish are drab in colour and merge with their surroundings.

Scorpionfishes possess spines in the dorsal, anal and ventral fins. (The spines on the head of some species are also thought to be venomous.) These spines contain several grooves running lengthwise which contain the venom glands. A protective integumentary sheath covers each spine and on puncturing the victim's flesh, this sheath folds back exposing the venom glands which rupture and release their contents. Generally the amount of venom entering the wound depends on the depth of penetration.

The above is not the case, however, for the stonefishes, which are closely related to scorpionfishes. While the spines are similarly grooved, the mechanism for introducing venom into the wound is much more efficient. Two large venom glands are attached to the spine, each with a duct running along a spinal groove which terminates near the tip of the spine. Slight pressure on the spine causes the contents of the glands to be expelled at the spine's tip. Thus a stonefish can inject the entire venom contents of a spine into the wound, even if the spine only enters a short distance. As with the scorpionfishes, the more spines involved, the more severe the sting.

Plate 4: Gurnard Perch

The following discussion deals with those scorpionfishes and stonefishes most often encountered in Western Australian seas:

Gurnard Perch (Plate 4)
Neosebastes pandus

This rocky reef species may be distinguished from other southern Western Australian scorpionfishes by its large, wing-like pectoral fins which extend back past the commencement of the anal fin. It is mottled brown to orange in colour with small, dark spots. It grows to a length of 40 centimetres. The Gurnard Perch has apparently been responsible for the death of one angler in Western Australia. The sting brought on a heart attack.

Distribution: southern half of Australia, from Victoria to the Houtman Abrolhos in Western Australia.

Red Scorpion Cod (Plate 5)
Scorpaena sumptuosa

An inhabitant of rocky reefs in the south-west of Western Australia, the Red Scorpion Cod, or Red Rock Cod as it is also known, is treated with respect by fishermen familiar with it. It is occasionally caught by anglers fishing for reef species, as its large mouth enables it to take snapper-sized baits. It may be distinguished from other closely related scorpionfishes in southern Western Australia by its overall reddish colour (see Plate 5) and 12 dorsal spines. Both the Gurnard Perch and Gulf Gurnard Perch possess 13 dorsal spines. The Red Scorpion Cod grows to about 40 centimetres in length.

Distribution: Western Australia only, from Albany to Shark Bay.

Gulf Gurnard Perch
Neosebastes panticus

Another reef-dwelling species from the southern half of Western Australia, this scorpionfish may be distinguished from the above two species by a combination of characteristics, including 13 dorsal spines, pectoral fins not extending to the commencement of the anal fin and a large black blotch on the dorsal fin between the third and eighth spines. It grows to a length of 28 centimetres.

Distribution: South Australia to the Houtman Abrolhos in Western Australia.

Marbled Scorpion Fish (Plate 6)
Parascorpaena picta

This fish is common in shallow waters around islands and inshore areas along our

Plate 5: Red Scorpion Cod

north-western coast. It is generally encountered when turning over rocks uncovered at low tide. It grows to 16 centimetres in length.

Distribution: Indo-West Pacific. It is found as far south as Shark Bay in Western Australia.

Fortescue (Plate 7)
Centropogon australis

This species is found in weed beds around coastal reefs in our southern waters. It grows to a length of 15 centimetres.

Distribution: southern half of Australia to Geraldton in Western Australia.

Plate 6: Marbled Scorpion Fish

Plate 7: Fortescue

Plate 8: Red Firefish

Red Firefish (Plate 8)
Pterois volitans

Also known as the Butterfly Cod, this species is easily recognised by its zebra-like colour pattern, very slender dorsal spines and large fan-shaped pectoral fins. This slowly swimming fish is usually found 'hovering' close to reefs or in caves. Its spines are extremely venomous and a sting, although not as dangerous as that of the stonefishes, is very severe. This species grows to a length of 38 centimetres.

Distribution: Indo-West Pacific. In Western Australia, it is found as far south as Rottnest Island.

Devil Fish (Fig. 6)
Gymnapistes marmoratus

Easily recognised by its scaleless skin, the Devil Fish or South Australian Cobbler is a nuisance to fishermen using nets along our southern coastline. The pain resulting from a sting from this weed-bed inhabitant is intense and far out of proportion to the size of the fish. It grows to about 20 centimetres in length.

Distribution: southern half of Australia to Fremantle in Western Australia.

Fig. 6: Devil Fish

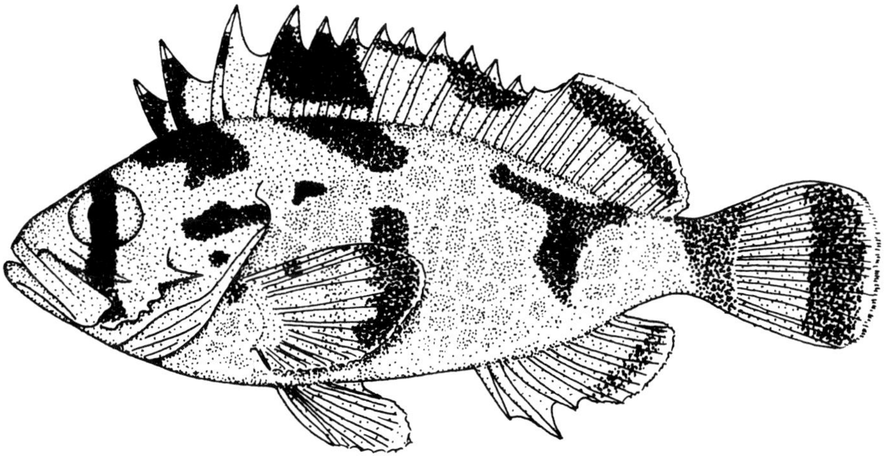

Common Stonefish (Fig. 7)
Synanceja horrida

The deadliest of all venomous fishes, the Common Stonefish is a sluggish animal which relies on camouflage to avoid detection. It is found in our more northern waters where it is occasionally caught by anglers or taken in craypots. The pain from a sting is usually excruciating and the victim must be restrained and immediately given first aid. This species has been responsible for a considerable number of deaths, mostly outside Australia. It may grow to a length of 60 centimetres.

Distribution: northern half of Australia to the Houtman Abrolhos in Western Australia. Also known from the Indonesian-Malaysian region.

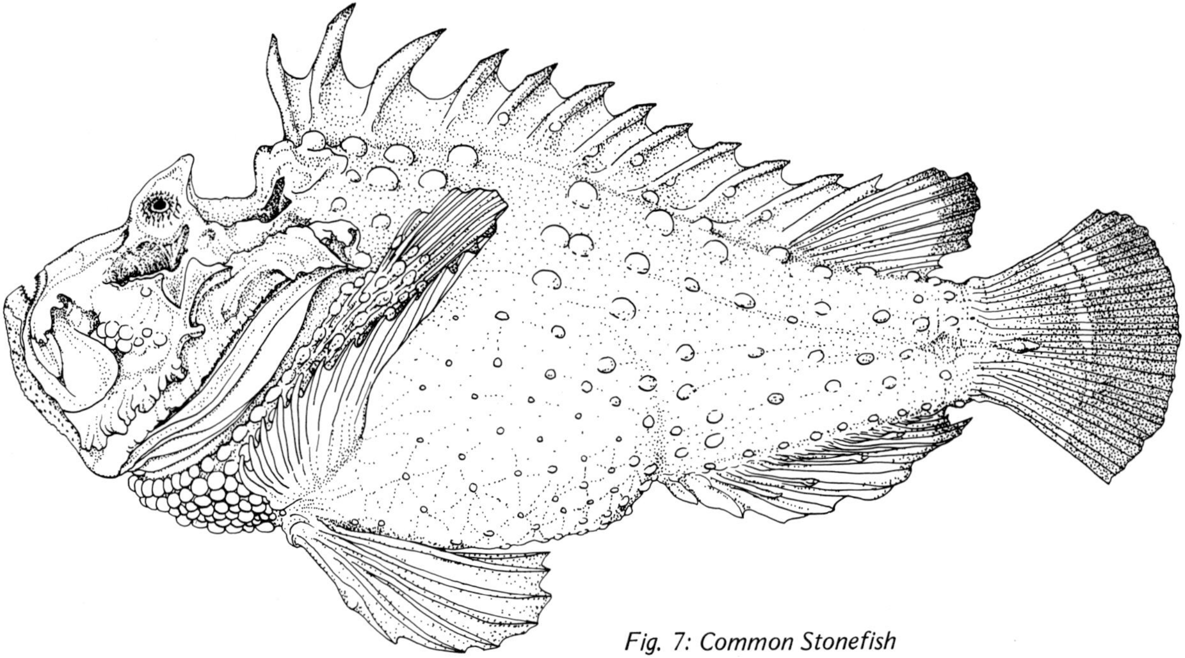

Fig. 7: Common Stonefish

Spottedfin Stonefish (Plate 9)
Inimicus sinensis

This species is possibly as deadly as the Common Stonefish. It is occasionally caught by trawlers operating in Shark Bay and Exmouth Gulf. It is easily distinguished from the Common Stonefish by the more slender dorsal spines and yellow spots on the inside face of the pectoral fin. It grows to a length of 33 centimetres.

Distribution: Indian Ocean and as far south as the Houtman Abrolhos in Western Australia.

Plate 9: Spottedfin Stonefish

Prevention, Symptoms and First Aid

The advice listed under these headings in the section on catfishes (see page 11) also generally applies for scorpionfishes and stonefishes. In addition, note the following:

- Divers should be very careful when placing their hands in reef holes or under ledges, as a number of scorpionfish stings have occurred from such actions.
- Many species of scorpionfish rest motionless on the bottom (see Plate 5). Never attempt to pick one up by hand.
- A victim of a stonefish sting should receive medical aid as soon as possible. An antivenene is available.

Stingrays

These rays are best known for the venomous spines, typically one or two, located on their tails. Their habit of lying motionless on the bottom, sometimes partly covered by sand, makes them particularly dangerous to waders and divers who may accidently tread or land on them. A stingray's first action in defence is to rear up its tail, thrusting the spine into the victim. These repeated thrusts can cause terrible wounds, with the serrated spine ripping the flesh on each withdrawal.

The venomous spine of stingrays possesses characteristics of both catfish and scorpionfish spines. The spine is generally a tapering structure, flattened and much larger than those of other venomous fishes. Along each lateral edge is a row of backward-directed barbs which enables the spine to tear a victim's flesh on withdrawal. In this way it facilitates absorption of the venom. On the spine's underside are two grooves running lengthwise which contain the venom glands. The whole structure is covered by an integumentary sheath which folds back to expose the venom glands on penetrating a victim's flesh.

Smooth Stingray (Plate 10)
Dasyatis brevicaudata

This species is considered the largest in the world, growing to a length of 4 metres. It is grey to brown in colour and may be distinguished from the closely related Black Stingray (*Dasyatis thetidis*) by its smooth back, the latter species possessing several longitudinal rows of large spiny tubercles on its back.

Distribution: both species are found in the southern half of Australia, also in South Africa.

Plate 10: Smooth Stingray

Eagle Ray (Plate 11)
Myliobatis australis

This ray is most often observed swimming in shallow water over sandy bottoms. At night it comes right up to the water's edge when searching the bottom for shellfish, upon which it feeds. Unlike most species of ray, it is rarely found resting on the bottom. It is very wary of humans and is dangerous only if encountered in areas of poor visibility, such as the surf zone on beaches. The Eagle Ray is characterised by its protruding head, pointed pectoral fins and blue cross-bands on the back. It grows to a width of about 1.2 metres.

Distribution: southern half of Australia to Lancelin in Western Australia.

Plate 11: Eagle Ray

Plate 12: Common Stingaree

Common Stingaree (Plate 12)
Urolophus testaceus

This ray is common in areas of reef and sand in southern Western Australia. It is frequently observed by divers resting motionless on the bottom (see Plate 12), and is particularly dangerous when covered by sand. An unsuspecting diver could be severely injured if he landed on a buried stingaree. Several other species of stingaree are found in Western Australian waters but this species is the most abundant. Stingarees differ from other stingrays in possessing a prominent caudal fin at the end of the tail, instead of the whip-like tail generally found on stingrays. The Common Stingaree grows to a length of 75 centimetres.

Distribution: southern half of Australia to Lancelin in Western Australia.

Coachwhip Ray (Plate 13)
Himantura uarnak

This stingray is common in mangrove areas and tidal flats along our north-western coastline. Its colour pattern is variable, ranging from plain sandy brown to a spotted or reticulate pattern (see Plate 13). It grows to a length of about 1.5 metres.

Distribution: Indo-West Pacific including northern Western Australia.

Blue-spotted Fantail Ray
Taeniura lymma

This species is common on coral reefs and sand flats along our north-western coastline. It can easily be identified by the combination of blue-spotted colour pattern, oval body shape and wide fold of skin along the lower surface of its tail. This ray grows to about one metre in length.

Plate 13: Coachwhip Ray

Blue-spotted Stingray
Amphotistius kuhlii

This species may be distinguished from the Blue-spotted Fantail Ray by the angular shape of its pectoral fins (rounded in the latter species). It is usually found on sandy bottoms close inshore along our north-western coastline. However, it has been trawled off the coast at depths to 60 metres. This ray grows to a length of about 40 centimetres.

Distribution: Indo-West Pacific including the northern half of Western Australia.

Prevention, Symptoms and First Aid

The advice for catfishes listed on page 11 also generally applies to stingrays. In addition, the following information is applicable:

- When walking on shallow sand or mud flats, shuffle the feet to frighten off any stingrays resting on the bottom.
- Divers should look for partly buried rays when choosing an area of sandy bottom to rest upon.
- If a wader treads on something which moves under his foot, he should immediately raise his foot. A stingray may not strike if he can make his escape quickly.
- Don't pick up a live stingray between the eyes. It may still be able to reach your hand with its spine.
- A stingray wound may be very large due to the slashing and ripping effect of the thrusting spine. In this case, bleeding must be controlled. This can almost always be done by covering the wound with clean absorbent material, such as a towel, and applying firm pressure. If

the blood flow from a limb cannot be controlled in this way (and this is rare) a constrictive bandage or tourniquet may be applied on the heart side of an injury to a limb, preferably over the upper arm or thigh. This tourniquet must be of wide material such as a wide bandage, not a rope, and must be applied tightly enough to stop the pulse. It should be left on for a maximum of 1½ hours, and after release should not be reapplied as this may cause permanent damage. However it is emphasized that most wounds other than those resulting from shark attacks can usually be controlled by the former method and a tourniquet is rarely necessary.

- If a sting is severe, the victim may be in extreme pain and possibly paralysed, so must be rescued to avoid drowning.

- The wound should be examined for foreign matter such as bits of the stingray's spine and integument, and any pieces removed.

- Medical treatment is imperative, as the risk of infection is high with stingray wounds.

- Symptoms may last from a few hours to several days. Some symptoms may persist for months.

- A few deaths have been reported from stingray wounds but these have generally occurred where the victim's chest was penetrated by the spine.

Spinefeet

The harmless-looking spinefeet are mostly schooling fishes. Some species possess bright colour patterns while others are drab (Plate 14). They all have venomous spines in the dorsal, anal and ventral fins and may be identified by the large number of anal spines (seven; most other fishes possess four or less anal spines). They have a variety of common names, including Rabbitfish, Happy Moments and Black Trevally. They are usually found around our northern coral reefs with some species infrequently reaching as far south as Rottnest Island.

The venomous spines of spinefeet are similar to those of scorpionfishes in both structure and mechanism of stinging (see page 12). However the reaction to a sting is generally not as severe as to that caused by a scorpion-fish.

One of the commoner species of spinefoot, *Siganus nebulosus,* is illustrated in Plate 14. It grows to a length of about 25 centimetres.

Plate 14: A spinefoot

Prevention, Symptoms and First Aid

The advice list for catfishes (see page 11) generally applies to spinefeet.

Miscellaneous Venomous Fishes

The following fishes are considered to possess venomous spines but are not dealt with here in detail for one or more of the following reasons:

- their venomous qualities have not been confirmed;
- they are rarely encountered in Western Australia;
- they are seldom responsible for stings.

Flatheads — e.g. *Platycephalus castelnaui* (Plate 15)
Stargazers — e.g. *Ichthyscopus barbatus*
Frogfishes — e.g. *Halophryne diemensis*
Goblin Fish — *Glyptauchen panduratus*
Pigfishes — e.g. *Congiopodus leucometapon*
Old Wife — *Enoplosus armatus*
Port Jackson Shark — *Heterodontus portus-jacksoni*
Butterfly Gurnards — e.g. *Chelidonichthys kumu*
Butterfishes — e.g. *Selenotoca multifasciata*
Red Velvet Fish — *Gnathanacanthus goetzeei*

It is also worth noting that any fish possessing stout spines is capable of producing a painful wound if not handled carefully. This pain has caused many people to incorrectly associate such wounds with stings. Nevertheless, any deep puncture should still be medically treated as a precaution against secondary infection.

Plate 15: Blue-spotted Flathead

Poisonous Fishes

The fishes dealt with in this category may contain poisons in their bodies and are therefore dangerous to eat. They can be divided into two groups:

- those which manufacture their own poison;
- those which acquire the poison from another source.

The first group contains pufferfishes, boxfishes and porcupinefishes, while the ciguateric fishes are placed in the second.

Plate 16: Ringed Pufferfish

Plate 17: Common Blowfish

Pufferfishes

Also known as blowfishes and toadfishes, pufferfishes get their common name because of their ability to inflate their abdomens by swallowing air or water. Many people, and countless pets, have died throughout the world from eating pufferfish flesh, although in Japan, with correct preparation, these fishes are considered a delicacy. In Western Australia at least four people have died as a result of eating these fishes.

The poison, called tetrodotoxin, is located in the internal organs and skin and is thought to be at its strongest concentration during the mating season (November to around April for the Common Blowfish). Careful removal of these parts of the fish may leave the flesh unaffected. However, this is not recommended. It has been reported that over 50 Japanese die every year from incorrect preparation of a pufferfish delicacy called Fugu. Captain Cook and some of his crew were severely poisoned after eating pufferfishes at New Caledonia. However, the only fatality was a pig which was given some scraps.

Common Blowfish (Plate 17)
Torquigener pleurogramma

This species is well known as a pest by anglers who fish in the Swan River estuary. However, some Western Australians regularly eat this fish after first skinning and gutting it. This practice cannot be considered safe, as the Common Blowfish has been responsible for the deaths of many pets in Australia. A study carried out in 1968 showed that while terrestrial animals are susceptible to the poison of this fish, marine animals such as seagulls and other species of fishes, appear to be immune. This species grows to about 22 centimetres in length.

Distribution: southern half of Australia (except Victoria and Tasmania) to Shark Bay in Western Australia.

Ringed Pufferfish (Plate 16)
Arothron armilla

This species is relatively common in coastal waters of southern Western Australia where it inhabits offshore reefs. It is easily identified by the blackish-brown ring around the base of the pectoral fin (Plate 16). It grows to a length of 26 centimetres.

Distribution: southern half of Australia to Green Head, north of Lancelin in Western Australia.

Silver Pufferfish (Plate 18)
Lagocephalus sceleratus

Also known as the Northwest Blowfish, this fish has quite a reputation. As well as being responsible for poisoning Captain Cook, it often attacks swimmers and divers, removing pieces of flesh and even toes. The powerful, parrot-like beak can easily bite through such things as snapper hooks and human bone. This species is more likely to be eaten than most other pufferfishes because of its large size (over 80 centimetres in length).

Distribution: Indo-West Pacific. In Australia, it ranges as far south as South Australia.

Plate 18: Silver Pufferfish

Boxfishes

These fishes are encased in a bony armour of fused scales, with openings for the mouth, eyes, gills and fins. Those that possess prominent spines over the eyes are commonly referred to as cowfishes. Some species are very colourful (see Plate 19). They are rarely eaten because of the bony armour, which makes them appear unattractive as a food fish. However, the poisonous quality of the internal organs is similar to that of pufferfishes and makes them potentially dangerous.

White-barred Boxfish (Plate 19)
Anoplocapros lenticularis

This reef-dwelling boxfish is common in coastal waters of the south-west of Western Australia. The orange and white colour pattern of the adult is its chief diagnostic feature (see Plate 19). This species grows to 25 centimetres in length.

Distribution: South Australia to Lancelin in Western Australia.

Plate 19: White-barred Boxfish

Porcupinefishes

These fishes are easily recognised by the large spines contained in the skin. These are capable of erection when the fish inflates itself. The internal organs contain a similar poison to that of pufferfishes, but due to the inedible appearance of porcupinefishes, few cases of poisoning have been reported.

Globe Fish (Plate 20)
Diodon nicthemerus

This species of porcupinefish inhabits reef and weed areas along our southern coastline. It is readily identified by four black bars on the sides and the yellow spines. Several different species occur in our northern waters. The Globe Fish grows to a length of about 40 centimetres.

Distribution: southern half of Australia to Lancelin in Western Australia.

Plate 20: Globe Fish

Prevention, Symptoms and First Aid

Prevention

Never eat the flesh of any fish that either is capable of inflating its abdomen or possesses a bony, box-like carapace. Also, such fishes should not be fed to pets.

Symptoms

- Within five to 30 minutes of eating flesh containing tetrodotoxin, muscular paralysis develops and may eventually lead to respiratory failure.
- The victim becomes weak and dizzy and may complain of curious sensations to the lips and tongue (swelling and numbness respectively).
- Vomiting almost always occurs.
- The victim may lapse into unconsciousness and unless medical treatment is received quickly, death could result.

First Aid

- Vomiting should be induced when the symptoms first become apparent, but not if the victim goes into respiratory failure.
- Artificial respiration should be given to the unconscious patient until medical assistance can be obtained.
- Hospitalization is imperative.

Plate 21: Chinaman Fish

Ciguatoxic Fishes

Ciguatoxic poisoning or ciguatera is an illness which follows the ingestion of various tropical reef fishes. The toxin appears to accumulate in the flesh of certain carnivorous species (i.e. flesh-eaters) such as rock cods, trevallies, sea perches, moray eels and sea pikes or barracudas. It is usually the larger and hence older individuals which contain sufficient toxin to be classed as poisonous. The toxin is thought to originate in blue-green algae which grows on the reef's surface. This is eaten by algal grazers such as surgeonfishes and parrotfishes, which in turn are preyed upon by the carnivorous species. The toxin is retained in the flesh of this latter group. Only very old individuals may contain enough toxin to be poisonous to humans.

There have been no reports of ciguatera from Western Australia, but many cases have been reported from Queensland and some tropical islands of the Pacific and Indian Oceans.

Ciguatoxic fishes are not found universally over large areas but are generally localized in small regions. Even then, not all large individuals of the one species in such an area may be poisonous. The level of toxin could be related to varying individual feeding habits of the species concerned.

The following species have been responsible for most reported cases of ciguatera in Australia:

Chinaman Fish (Plate 21)
Symphorus nematophorus

This species belongs to the sea perch family Lutjanidae. It has been responsible for several cases of poisoning in Queensland. Basic colouring is reddish, usually with pale vertical cross-bars of irregular shape on the body, and scattered dark spots along the back. Young individuals are brighter in colour, with a series of blue wavy lines running lengthwise along the body. They also possess several filament-like rays in the dorsal fin which disappear with age. This fish grows to a length of about one metre.

Distribution: northern half of Australia. Its southern limit in Western Australia is Shark Bay. Also found in the Indonesian-Malaysian archipelago.

Red Bass (Plate 22)
Lutjanus bohar

This fish is also a member of the sea perch family and is well known for its ciguateric qualities. It has been reported as causing ciguatera from various areas in the Indo-West

Pacific, including Queensland. Like the above species, however, it is regularly eaten in Western Australia where no cases of ciguatera have been reported. The basic colour of this species is red with the upper portion of its head and body a much darker red. Small individuals possess two round white blotches on their upper sides, one below the spines of the dorsal fin and the other below the rays. The Red Bass grows to a length of about one metre.

Distribution: Indo-West Pacific. In Western Australia its southern limit is Shark Bay.

Plate 22: Red Bass

Prevention, Symptoms and First Aid

Prevention

- Avoid eating large tropical reef-dwelling carnivorous fishes.
- If such a fish is caught and eaten without any apparent side-effects, do not keep any left-overs for further meals. The poison is lost very slowly from the body and while the amount ingested in one meal may not be toxic, additional meals could produce ciguatera by raising the level of poison in the body.

Symptoms

Symptoms of ciguatera usually start from two to 12 hours after the meal has been eaten but may not appear until after 30 hours. They include the following:

- A general weakness and dull aching in the limbs.
- Tingling in the hands, feet, lips and mouth.
- Itching of the skin in the palms of the hands and on the soles of the feet.
- Vomiting and diarrhoea.
- Headache, toothache and tightness of the chest.
- Reversal of perception to hot and cold – cold drinks appear hot and vice versa.
- Convulsions, vision irregularities, and paralysis.

Many deaths have resulted from ciguatera but these have occurred in areas outside Australia.

First Aid

- Induce vomiting by inserting fingers down the throat.
- Seek medical assistance.

Food poisoning is sometimes diagnosed as ciguatera but may be distinguished from it by the absence of tingling in the hands, feet, lips and mouth which is a diagnostic feature of the latter condition. While the gastro-intestinal symptoms of these two illnesses may be similar, generally the neural symptoms are more severe for ciguatera. Nevertheless, food poisoning can completely disable the victim and hospitalization may be required. Poisoning can, however, be prevented by following these points:

- Never allow a fish destined for the meal table to deteriorate. If caught on a hot day, it should be gutted and cleaned as soon as possible after capture and placed in a cool area out of the sun. Alternatively, it may be kept alive in some form of receptacle until the day's fishing is complete. However, placing dead fish in such a receptacle will not prevent bacterial attack, especially under tropical conditions. In such a case, bacteria in the water as well as those in the fish itself play a part in the flesh's deterioration.
- Fish flesh deteriorates at a much faster rate than meat, with some species deteriorating more rapidly than others. Therefore, always store fish under freezer conditions, never in the main compartment of a refrigerator unless it is to be eaten the same day.

Aggressive Fishes

This category usually receives most attention from the news media. People are therefore more aware of the dangers involved. Fishes responsible for attacks on humans are generally large predators, those flesh-eating species which prey upon other fishes. Sharks have been responsible for most attacks, although gropers, moray eels and barracudas have accounted for some. However, attacks by the latter three have usually occurred when these fishes have tried to relieve spearfishermen of their catches, or when they themselves have been speared and then turned on those responsible.

Sharks have certainly earned their reputation as man-eaters. Almost 350 attacks have been reported for Australian waters since 1791, although only 20 of these occurred in Western Australia. However, the species of sharks reponsible for these have been positively identified on relatively few occasions. Nevertheless, the potentially dangerous sharks are well known.

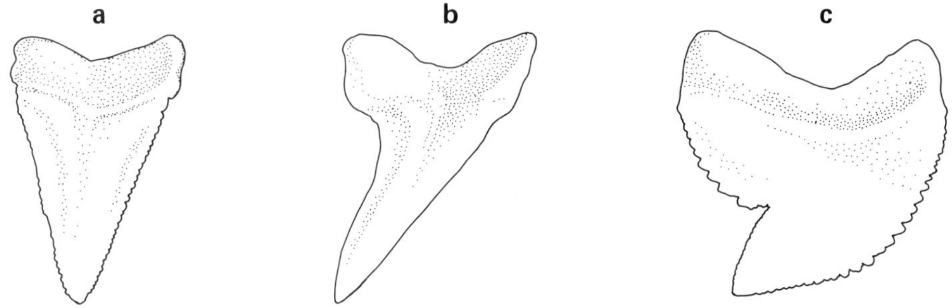

Fig. 8: Characteristic shapes of some shark teeth (a) White Pointer (b) Mako Shark (c) Tiger Shark

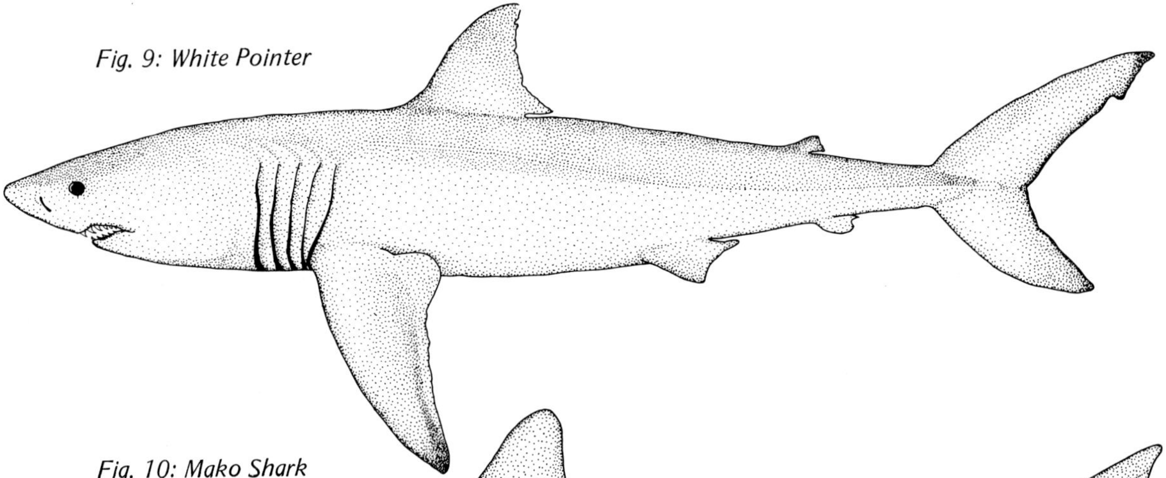

Fig. 9: White Pointer

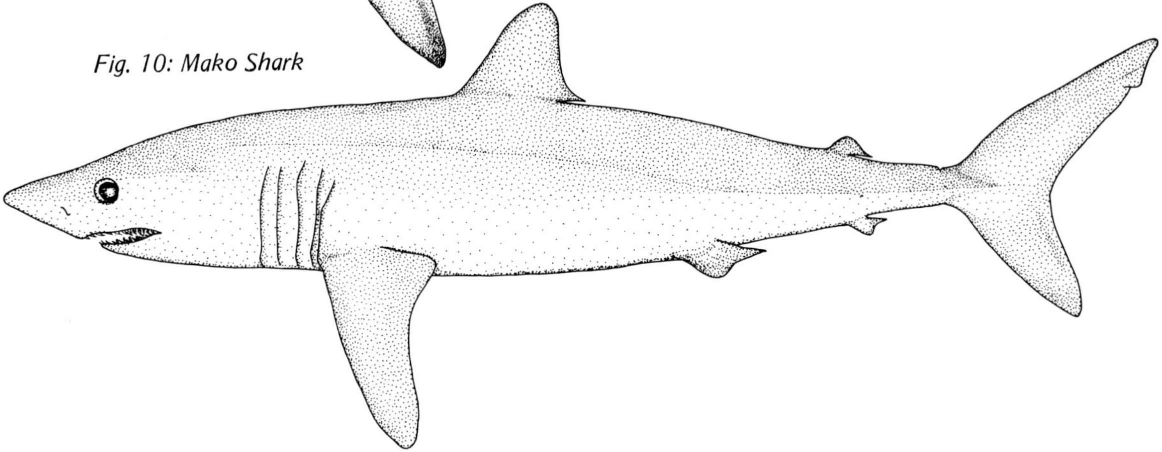

Fig. 10: Mako Shark

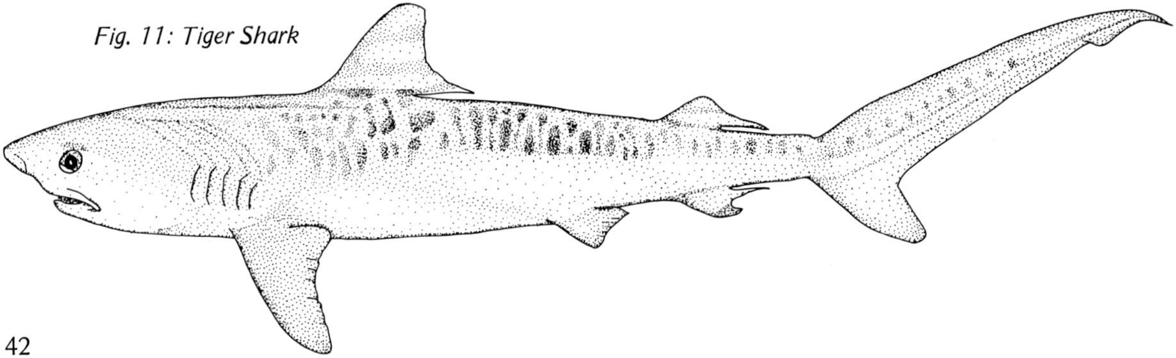

Fig. 11: Tiger Shark

Sharks

White Pointer (Fig. 9)
Carcharodon carcharias

The infamous White Pointer is a deep sea shark, preferring oceanic conditions. It occasionally comes close to shore, especially when attracted by such things as whale carcasses. This happened frequently at Albany when the whaling station was in operation. The White Pointer is occasionally taken by professional shark fishermen operating offshore. Abalone divers working on our south coast have had some frightening encounters with this shark.

This species is probably the most dangerous shark in the world. Its great size (to about 8 metres in length) and aggressive nature produce both fear and respect in every person who encounters one. It is readily identified by a combination of characteristics including a cone-shaped snout, lower lobe of the caudal fin only a little shorter than the upper lobe (see Fig. 9) and the teeth in both jaws large and broadly triangular (Fig. 8a). Its colour is generally greyish with a dirty white ventral surface.

Distribution: world-wide, preferring cooler temperate seas. In Western Australia it has been frequently recorded from the southern coastline and infrequently to as far north as the Dampier Archipelago.

Mako Shark (Fig. 10)
Isurus oxyrinchus

This oceanic shark occasionally comes close to shore. It is a very aggressive species which fortunately confines most of its attacks to boats. It is closely related to the White Pointer, but may be distinguished by its narrow, dagger-like teeth (Fig. 8b). The Mako Shark's colour is dark blue to bluish-grey. It grows to a length of about 3.6 metres.

Distribution: world-wide.

Tiger Shark (Fig. 11)
Galeocerdo cuvieri

This shark is at home in both oceanic and inshore waters, even occasionally being found under estuarine conditions. It is very aggressive, and large individuals (3.6 metres and over) are extremely dangerous. It is a scavenger, swallowing almost anything that comes its way, including ships' garbage, sea-birds, turtles, sea-snakes, stonefish, stingrays, blowfish and even seaweed. The extremely blunt snout and characteristic cockscomb-like teeth (see Fig. 8c) help distinguish this shark. Large individuals are greyish-brown in colour, while smaller examples possess a tiger-like pattern of bars on the body. It grows to a length of 6.4 metres.

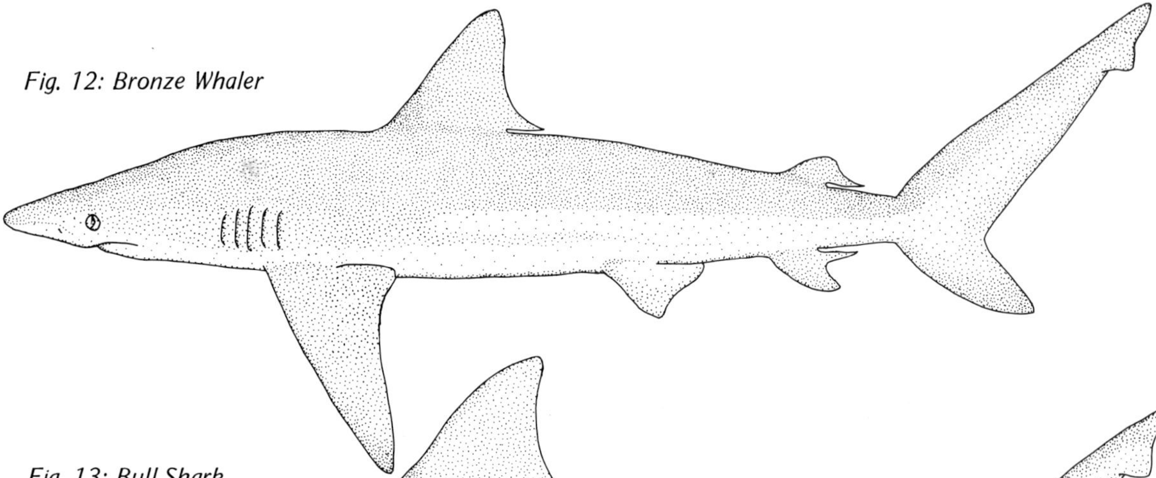

Fig. 12: Bronze Whaler

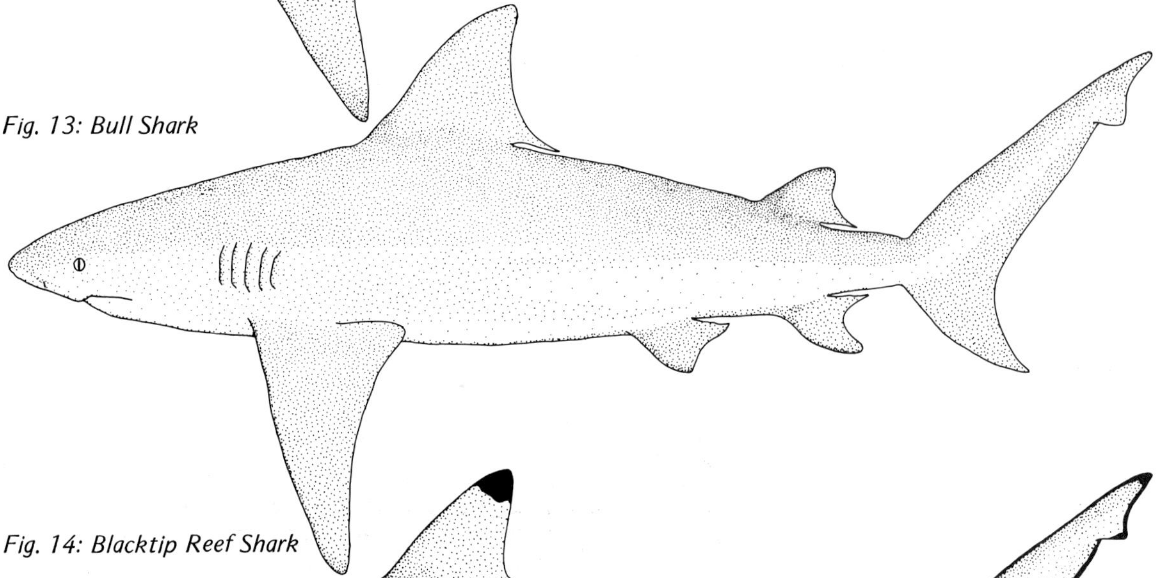

Fig. 13: Bull Shark

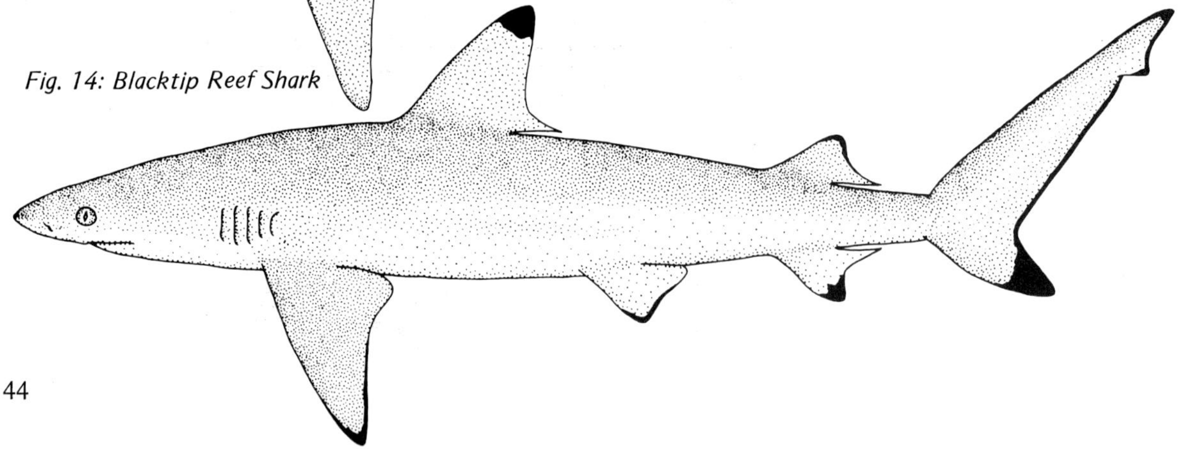

Fig. 14: Blacktip Reef Shark

Distribution: world-wide, preferring tropical seas. In Australia it is found as far south as South Australia.

Whaler Sharks

Twelve species of whaler sharks are found in Western Australia, including the Bronze Whaler (*Carcharhinus brachyurus*), Dusky Shark (*Carcharhinus obscurus*), Bull Shark or Swan River Whaler (*Carcharhinus leucas*), Blacktip Reef Shark (*Carcharhinus melanopterus*) and Sandbar Shark (*Carcharhinus plumbeus*). All are considered dangerous to man. They are sleek in appearance, with rounded to pointed snouts, long pectoral fins, caudal fins with short lower lobes and short gill-slits. Colouration ranges from grey to bronze, with some species possessing black markings on the fins. Unlike the other species of shark discussed, whaler sharks often travel in schools. Most species are difficult to identify. For this reason only three of the more important species are discussed here.

Bronze Whaler (Fig. 12)
Carcharhinus brachyurus

This is the common whaler shark of southern Western Australia. The common name of this species is derived from its colouration which is bronze to a brownish-grey. It may be distinguished from the closely related Dusky Shark by the absence of a skin ridge on the back between the two dorsal fins (present in the latter species). The Bronze Whaler grows to a length of three metres.

Distribution: world-wide, preferring cooler waters.

Bull Shark (Fig. 13)
Carcharhinus leucas

This shark is also known as the Swan River Whaler, as it has been frequently reported and caught in the Swan River. Although it is a marine species, it is able to tolerate estuarine and freshwater conditions and has been responsible for many attacks in rivers around the world. In Western Australia, attacks in the Swan River and at Broome have been attributed to this species. It has been reported from 320 kilometres up the Fitzroy River in the Kimberley region of Western Australia. It grows to 3.6 metres in length.

Distribution: world-wide, preferring warmer waters.

Blacktip Reef Shark (Fig. 14)
Carcharhinus melanopterus

The coral reefs and some mangrove areas of our northern waters abound with this species of shark. It prefers shallow water and although wary of divers, has attacked waders in depths of less than 30 centimetres. It is a small shark, growing to a length of 1.8 metres. The bronzy grey colour of its body and generally prominent black tips to all fins make it easy to identify.

Distribution: Indo-West Pacific. In Western Australia its southernmost limit is in the region of North West Cape.

Wobbegongs and Sawfishes

Although not normally aggressive, the wobbegongs and sawfishes are worthy of mention. The old saying, 'Let sleeping dogs lie', is certainly appropriate for these sluggish bottom-dwelling animals.

Wobbegongs, or carpet sharks as they are also known, possess a mottled colour pattern that helps them merge with the surroundings. They usually shelter during the day in caves or amongst seaweed, occasionally venturing into the open. Numerous attacks have been blamed on these sharks, although most have been provoked. A wobbegong resting in shallow water does not take kindly to being stepped on, and detests having its tail grabbed. They have always been easy targets for spearfishermen, but a poorly-speared wobbegong can be more than a handful for the unfortunate diver. Their fang-like teeth have caused serious wounds; a fisherman in New South Wales had his foot ripped off after accidentally treading on a wobbegong. In Western Australia we have five species of wobbegongs. These are easily separated from other sharks by the numerous tentacles around the mouth. The most common species is possibly the Spotted Wobbegong *Orectolobus maculatus,* which grows to a length of 3.5 metres.

Sawfishes belong to one of the families of rays. They are closely related to the shovelnose rays but differ in possessing a greatly elongated snout armed along each edge with a row of teeth in the form of a double-edged saw. This saw is about one quarter the length of the animal. Reports from India indicate that sawfishes have attacked humans, the slashing saw inflicting enormous wounds. However, these attacks were most likely provoked by the capture or attempted capture of these normally harmless animals. There are three species of sawfish in Western Australia, the best known being the Green Sawfish *Pristis zijsron.* It grows to a length of 7.5 metres with a saw of 1.8 metres. Its range includes the northern half of Australia to as far south as Shark Bay in Western Australia, and it is at home under both marine and freshwater conditions.

Prevention

Some authorities give so much advice that if all was heeded nobody would ever enter the water. However, some recommendations that should be followed are listed below:

- Always swim or dive with a companion, not because it affords a 50 per cent chance of the other person being attacked, but so that immediate assistance may be given.
- Blood attracts and excites sharks, so stay out of the water, or leave it, if you have an injury that is bleeding.
- Avoid swimming in dirty water. Poor visibility may lead to a shark mistaking you for a large, struggling fish.
- Never swim in an area where fish are being cleaned.
- A struggling fish emits vibrations that sharks can pick up from a considerable distance. Remove speared fish from the water as soon as possible and never attach them to your body. It is realistic to say that if you are frightened of sharks, don't take up spearfishing.
- Don't panic if you see a shark. Most are wary of humans and will generally swim off when approached. Move slowly towards land or nearby boat, keeping an eye on the shark (swimmers should hold their breaths and place their heads below the surface – a shark is more wary if it knows it can be seen). Some sharks like to test a proposed meal by bumping it with their snouts. They should be fended off with such things as the rear end of a speargun or a gloved hand. A sharp knock on the snout is usually enough to discourage an inquisitive shark. A bare hand or foot should only be used as a last resort as shark skin is very rough and may cause serious abrasions and bleeding.

First Aid

- In the event of an attack, every effort should be made to control bleeding, even while the victim is still in the water. A tourniquet should be applied above the wound if on a limb (see pages 24 & 25) or packing pressed against the wound if it is on the trunk. This prevents loss of blood, and also helps to reassure the victim. Both shock and loss of blood can lead to the victim's death.
- The victim should be carefully taken from the water and laid on the beach or boat-deck in a position with the legs higher than the head. Handling and movement should be as gentle as possible and kept to a minimum. If the victim has to be brought ashore by boat, see that this is done carefully and gently. The victim should not be moved again until medical help arrives. Many shark victims have died from the shock of being rushed to hospital by the rescuers.
- Reassure the victim (if conscious) and cover lightly with a dry towel or similar article. Nothing should be given by mouth.

Further Reading

Dangerous Marine Animals of the Indo-Pacific Region by Dr C. Edmonds (1975). Wedneil Publications, Newport, Victoria.

Australian Venomous and Poisonous Fishes by R.V. Southcott (1975). R.V. Southcott, Adelaide, South Australia.

Acknowledgements

I wish to thank Gaye Roberts whose excellent line drawings illustrate many of the fishes included in this booklet. Gerald R. Allen kindly provided a transparency of the Spotted-fin Stonefish, while those of the Chinaman Fish and Red Bass were generously supplied by Roger C. Steene. To Ann R. Fernandez of the Museum's Publications Department I am particularly grateful for assistance with the booklet's preparation. Finally, I would like to express my thanks to the many people who took the time to inform me of their encounters with our dangerous fishes.

Advice on first aid follows recommendations by the Australian Resuscitation Council.